Hamilton Wood Type

A HISTORY IN HEADLINES

by
Bill Moran
Robert Style
Dennis Ichiyama
and Richard Zauft

with photos by David Ray Carson

This book is dedicated to the memory of Rob Roy Kelly (1925-2004) - designer, educator and historian - whose passionate exploration of the use of wood type reawakened interest in this centuries-old art form. His seminal book, *American Wood Type, 1828-1900: Notes on the Evolution of Decorated and Large Types and Comments on Related Trades of the Period*, continues to inspire designers and printers to investigate and apply wood type's unique aesthetic.

Preface

In 1880, a young Two Rivers entrepreneur named James Edward Hamilton fulfilled his dream of founding his own company. His product, wood type, became the printers' product of choice across the United States.

Even after phototypesetting was created, wood type remained the standard for headlines and posters into the 1980's. In the late 1890's, Hamilton diversified into wood cabinets to store the type, and in 1904, entered into the laboratory casework business. Today Fisher Hamilton is the world's largest manufacturer of laboratory furniture and airflow products.
It continues to have an economic and social impact on the local community as the largest employer in Manitowoc County.

Fisher Hamilton is pleased to support the Museum on the location where the legacy began and help in its efforts to preserve a significant chapter in the story of American printing.

J. Michael Brown
Fisher Hamilton
President and CEO

The original Hamilton Manufacturing Company smokestack stands on the Fisher Hamilton site as a reminder of the company's rich history.

Table of Contents

This book was published with the assistance of grants from Fisher Hamilton,
The University of Minnesota Design Institute, and the Two Rivers Historical Society.

ISBN 0-9723927-1-8

Design by Bill Moran, Mary Maguire and Martine Forte
Primary photography by David Ray Carson
Photography on pages ii, iii, 3, 4, 5, 18, 20, 21, 23, 24, 25, 26, 27, 28, 32, 33, 35, 57
copyright 2004 David Ray Carson
Printed on Environment® papers from Neenah

Foreword

So there I was in Norway, one spectacular and sunny autumn afternoon. Sitting in my friends backyard, I was marveling at all the trees and the quilting of their changing colors, leaves shimmering in brilliant hues painting one last gasp before winter.

Then my Norwegian friend said that his father had worked on a cruise line back and forth across the Atlantic. He said when he was a boy his father always brought with him the latest modern toys from America, and how he never liked them.

I asked him why and he said that his toys were too modern and complicated, or they always needed new batteries, which were expensive and hard to find. He said "the wooden toys from the other boys were more reliable."

Being a letterpress poster printer I knew what he meant. At Hatch Show Print my staff and I work and play with wood type everyday. We're familiar with their individual little scars and dents and cherished flaws as a result of decades of use. There's an innate humanity, an organic quality to working with the countless pieces of type in our shop day after day. They've lasted longer than most of the entertainers who paid for their services.

As we celebrate our 125th anniversary in 2004, we honor the heritage of the shop, and the wood type that still makes it possible for us to complete hundreds of jobs annually. There's a story to be told in the heart of every piece of that wood.

Jim Sherraden (right) watching wood type cutter Norb Brylski prepare templates for routing.

Such is the Hamilton story. And it needs to be told because little by little designers and printers are yearning to touch something real, thirsty to evolve from the computer. An interest in movable type is at hand.

As a Luddite, I'm glad to see this movement rustling these days like those leaves in that autumn Norwegian forest. Where once more the brilliant hues of nature catch their breath, landing lightly in the eye, and color once more bursts from their grain.

Jim Sherraden, Hatch Show Print, Nashville Tennessee April 2004

Introduction

The Hamilton Wood Type and Printing Museum is home to one of the largest collections of wood type and associated material in the United States. Founded to honor the innovative work of Ed Hamilton and the exceptional craftsmanship of the many employees of the Hamilton Company. The Museum is a testament to the importance of manufactured wood type to the history of printing in the United States.

An early version of the Hamilton Wood Type logo.

Today the museum is home to 1.5 million pieces of wood type. Visitors can see first hand each step of the painstaking process, from the industrial routers spinning at 50,000 revolutions per minute to the exquisite care needed to carve serifs into a single piece of type. Since the museum opened its doors in 1999, it has hosted printmakers, wood engravers, typographers and historians as part of its visiting artist series.

Workers outside the Hamilton factory at the turn of the century with Ed Hamilton in the doorway (left). Entrance to the Hamilton Museum (above) and a sampling of Gothic and Tuscan type (right).

One of the early Hamilton plants still in use today (right) and a Hamilton retiree volunteer putting the finishing touches on a new piece of type (far right).

The robust growth in demand for newspapers and printed advertisements in the United States after the Civil War created an unprecedented market for large, inexpensive type faces. James Edward Hamilton began making wood type near the turn of the century and built his company to take advantage of the emerging market of printing supplies. The Museum stands near the site of one of Hamilton's first factories, linking Fisher Hamilton with the spirit of its founder.

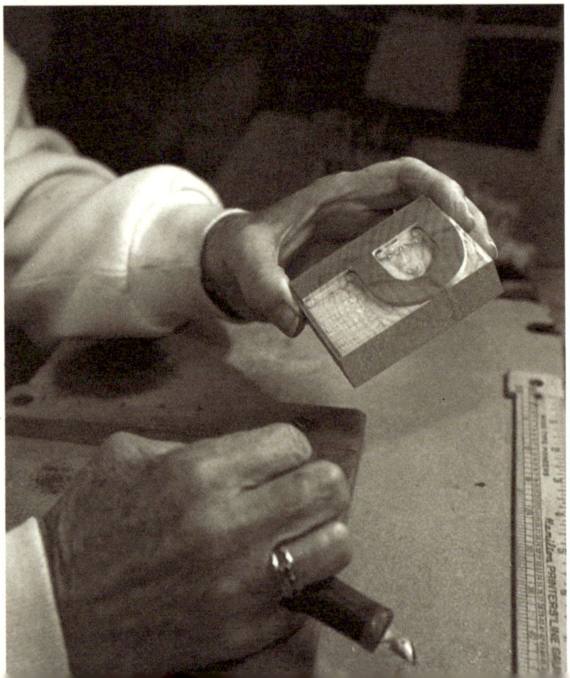

Hamilton Wood Type and Printing Museum is an excellent place to learn about the manufacture and use of wood type, the evolution of printing from the turn of the century to the modern era, and the emergence of the printing industry as one of the engines that drove the Industrial Revolution of the late nineteenth and early twentieth centuries.

The Museum's printing facility is used by printmakers to work with original wood type blocks (top) and its collection is an important resource for designers, historians and typographers worldwide (above).

Chapter 1
BEGINNINGS ON THE FRONTIER

James Edward (Ed) Hamilton was born on May 19, 1852 in Two Rivers, Wisconsin. His father, Henry Hamilton, had moved from New York to Two Rivers to take a position as a druggist in a store owned by Hezekiah "Deacon" Smith, a successful local businessman. While working in Deacon Smith's store Henry met the Deacon's daughter, Diantha Smith, who had just returned to Wisconsin after attending finishing school. Henry courted Diantha and eventually married her. Shortly afterward, he became involved in a number of business ventures with the assistance of his father-in-law.

James Edward Hamilton

Despite the Deacon's substantial influence in the Two Rivers' business community, none of Henry Hamilton's ventures were successful. In the 1850's, Henry became interested in the political life of the newly established State of Wisconsin culminating in his election to the Wisconsin Legislature in 1857. A rift developed between the Deacon and his son-in-law over Henry's foray into politics, a pursuit the Deacon felt was a waste of time for a serious businessman. The conflict with his father-in-law led Henry to move his family to Waucousta (now Greenbush) Wisconsin where, with his brother-in-law, he opened a combination gristmill/sawmill and a general store.

After the outbreak of the Civil War, Henry joined the Wisconsin Volunteer Army. Ed's mother moved the family back to Two Rivers to be near her parents. It was there that the family received news of Henry's death in Tennessee in 1864.

Hezekiah "Deacon" Smith

Henry Hamilton

The Hamilton family – then consisting of twelve-year-old Ed, his newly widowed mother and three younger siblings – moved to Lockport, New York where Henry's two brothers lived. Young Ed was enrolled in high school, which he attended for two years until the dire financial situation of his family required him to go to work. The family got by with the help of Henry's brothers and the meager earnings of the fourteen-year-old Ed, but Diantha had begun missing her friends and family back in Wisconsin. So, in 1868, the family packed up what little they had and returned to the frontier environment of Two Rivers.

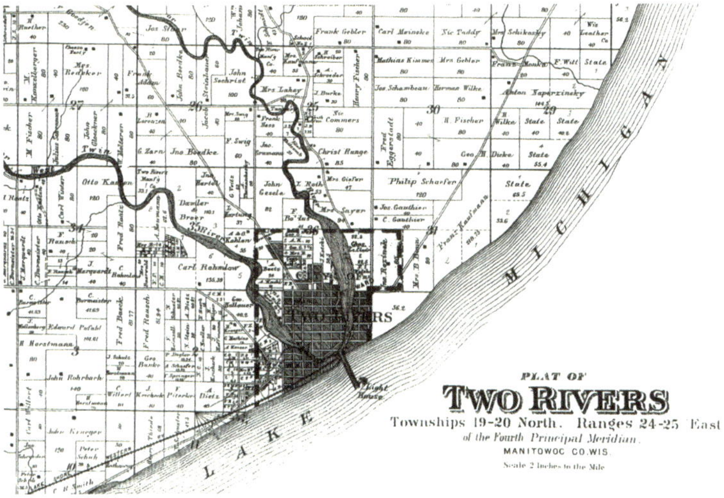

Platte map of Two Rivers in the second half of the nineteenth century.

The Port of Two Rivers in 1858.

Timber from northern Wisconsin at the convergence of the Twin Rivers in 1886.

Lumberjacks harvesting Wisconsin hardwood.

Lumber and Wood Type

In the second half of the nineteenth century, Two Rivers was a bustling lumber town dominated by the activities of the local sawmill and other timber-related businesses. Logs cut from the interior of the northern Wisconsin wilderness were sent down the Twin Rivers to Lake Michigan, where they were milled for expanding markets in the Midwest. The town was filled with burly lumber men who plied their trade with cant hooks, chains and axes guiding the rough logs along the river toward the huff and buzz of the sawmill.

A vigorous, wiry youth of seventeen, Ed Hamilton had developed a love for the outdoor life during his earlier days in Two Rivers. He loved being able to hunt and fish right out his back door, often sneaking off to tramp through the woods or sit pole-in-hand on the riverbank. He also enjoyed the bustle of the lumber crews that were a constant fixture in town. As Deacon Smith's grandson, he was able to walk easily among the lumberjacks, listening to their stories and learning to run across the logs as they were steered down the river to the sawmill - a risky pastime and that Ed took part in over the strong objections of his mother.

Lumber harvesting crews worked through the winter when the woods were easier to access.

In addition to riding logs down the river with the lumberjacks, young Ed Hamilton also hung around the sawmill and became fascinated with the machinery used convert the rough timber into useful wood stock. Not only did Ed enjoy watching the saws and lathes buzz through the rough logs, he also became fascinated with the business aspects of milling and the manufacture of wood products. The seeds of his entrepreneurial spirit were planted amid the mud, steam and sawdust of mills and factories of the booming, post-war Two Rivers.

Already a skilled woodworker, having honed his skills making toys and kites for local children, Ed got a job tending a clothespin lathe at a local pail factory. Hamilton took to the job immediately, challenging himself to figure out ways to step up his production and improve the quality of his work. His time in the factory also allowed him to pursue his interest in the machinery of production, eventually leading him to be a volunteer helper around the engine room of the factory. There he developed an engineer's eye, learning the basics of steam power and acquiring the mechanical skill that allowed him to maintain and repair equipment of all kinds. Engineering on the frontier was a hands-on endeavor and the same skills that made Ed Hamilton an excellent woodworker helped him excel in his role as a mechanic and machine tool designer.

Ed's ability to run complicated machinery and his enthusiasm for work led his uncle to put him in charge of a brick factory when he was only nineteen. Although the need for brick in the region was not sufficient to sustain the business, Ed was able to use the factory's facilities to start his first business - a small concern that made wall brackets and other odds and ends of ornamental furniture for local clients. Ed's skill with a scrollsaw kept him in business for a time, but the venture eventually folded. He then went on to run pile driving equipment on the Sturgeon Bay Canal until he caught "gold rush fever" and went to South Dakota to seek his fortune with some other like-minded young men in 1876.

Advertisement for wood goods from the Manitowoc County Chronicle in 1880.

Diantha Smith's home where Ed Hamilton made his first pieces of wood type.

A New Beginning

Upon his return from the Dakotas a year later - broke but with a newfound resolve to make his way in the business world - Ed went back to work in the chair and pail factory.

Shortly after his return to Two Rivers, he met and fell in love with the town's new schoolteacher, Etta Shove. Unique by frontier standards, Miss Shove was college-educated and independent-minded, making her a good counterpoint to Deacon Smith's rigid, traditional worldview. The couple was married in 1880 and took up residence in Ed's mother's home in town.

As the newlyweds began their life together, they were presented with an opportunity that would eventually change their lives and the economic life of Two Rivers. As with many events that lead to the development of successful businesses, it all started with a seemingly simple request.

William Nash, the publisher of the Two Rivers Chronicle, had agreed to make some large posters for the "Grand Ball," a dance to be held in town. As he began laying out the job, Nash realized he would need type of a style and size not available in his small print shop. Since there was not time to order the type from his East Coast suppliers, Nash began looking for local alternatives. Having heard of young Ed Hamilton's skill with a scrollsaw, Nash approached the him to see if there was a way to make the large type he need out of blank wood stock with locally available tools. The two men discussed the general specifications of wood type and Nash drew a rough approximation of how he wanted the type to look. Hamilton devised a method of cutting the required type out of holly wood with his foot-powered scroll saw and mounted the pattern onto softer wood for use in the printing press.

The posters were a great success. Nash was so pleased with the product that he ordered four or five more fonts of display type for his shop, and the process for manufacturing "Holly Wood Type" was born.

Around that time Ed's brothers George and Henry had gotten into the newspaper business in the small Minnesota town of Detroit. Excited by the response his type had gotten from Nash, Ed sent some samples to his brothers, asked them to do some printing with it and tell him what they thought. The brothers responded that the product was of top quality and, in return for the new type, they sent Ed printed packaging labels that read "This is Holly Wood Type. If you are interested, write J. E. Hamilton, Two Rivers, Wisconsin, for more details." Henry suggested that Ed attach the labels to packages with types he cut and test the market for his type by sending samples to newspapers around the Midwest.

Orders for Holly Wood Type began arriving at the Hamilton household shortly thereafter. The *Green Bay Gazette* requested $2.50 worth of type – Hamilton Holly Wood Type Co's. first official order. This was followed closely by a $12.00 order from a newspaper in Chippewa Falls, Wisconsin. Buoyed by the success of his type making efforts, Ed quit his job at the chair and pail factory, set up his scrollsaw in a back room of his mother's house, and started making type full-time.

The "Grand Ball" logotype, Hamilton's first piece of Holly Wood Type.

The Manitowoc County Chronicle.

WILLIAM F. NASH. INDEPENDENT. PROPRIET

VOLUME IX. TWO RIVERS, MANITOWOC CO., WISCONSIN, TUESDAY, DECEMBER 7, 1880. NUMBER

GOSSIP OF THE DAY. | NEWS OF THE WEEK. | OFFICIAL DOCUMENTS. | WISCONSIN NEW

Banner from the Manitowoc County Chronicle, circa 1880.

Uppercase Antique face wood type and its printed image (at right).

Entries in a ledger from the early days of the J.E. Hamilton Company.

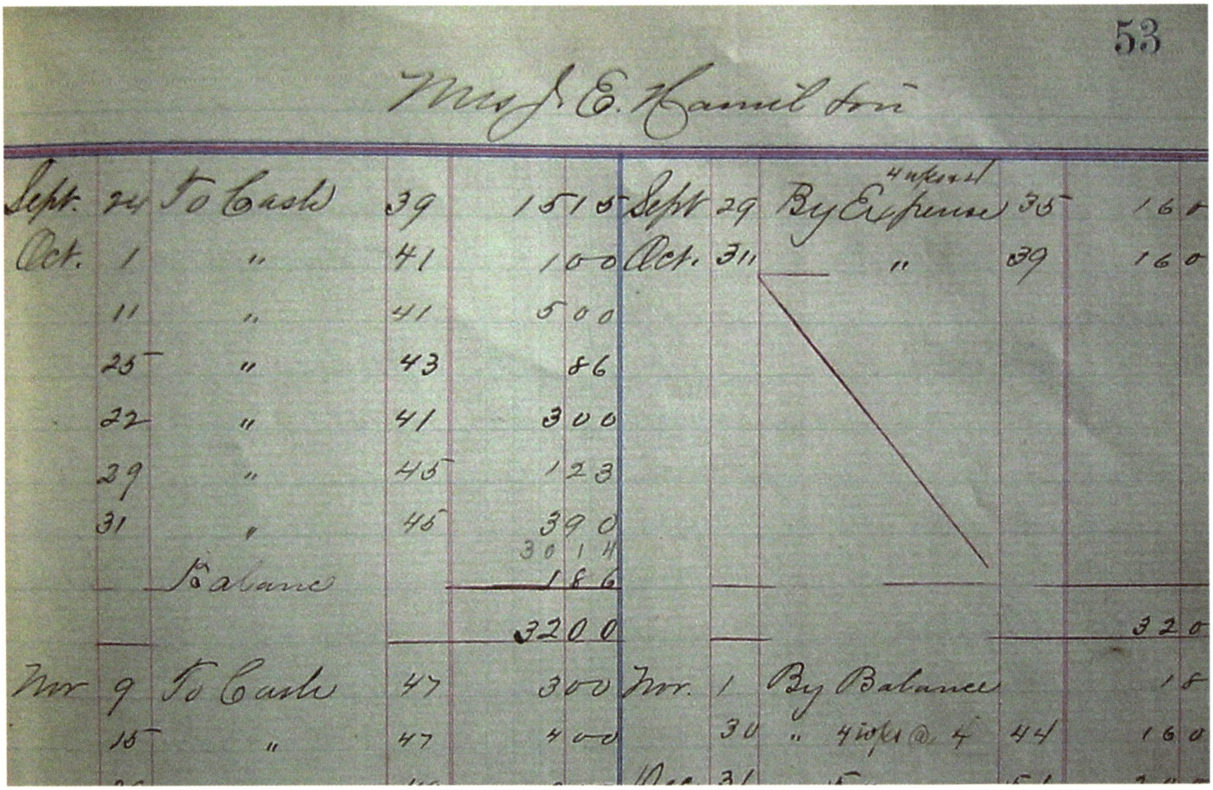

53

Mrs J. E. Hamilton

Sept.	24	To Cash	39	15 15	Sept	29	By Expense	35	16 0
Oct.	1	"	41	1 00	Oct.	31	"	39	16 0
	11	"	41	5 00					
	25	"	43	86					
	22	"	41	3 00					
	29	"	45	1 23					
	31	"	46	3 90					
				30 14					
		To Balance		1 86					
				32 00					32 0
Nov	9	To Cash	47	3 00	Nov.	1	By Balance		1 8
	15	"	47	4 00		30	" [illegible] @ 4	44	16 0

ABCDEFGHIJKLMN
OPQRSTUVWXYZ

With the help of his wife Etta, the company's first bookkeeper and cashier, Hamilton Holly Wood Type Co. began making and distributing wood type in 1880.

Like all new business owners, the Hamiltons had a lot to learn about running a successful enterprise. Deacon Smith was helpful in this regard, helping find partners to bring much needed capital to his grandson's business. The company was officially incorporated in 1889.

Over time, the Holly Wood Type manufacturing process was abandoned in favor of more durable end-grain type. As a result of Ed Hamilton's love of woodworking, his business sense, and his commitment to producing high quality products at fair prices, the Hamilton Manufacturing Company grew to become a leader in the production and distribution of wood type in the United States by the turn of the century.

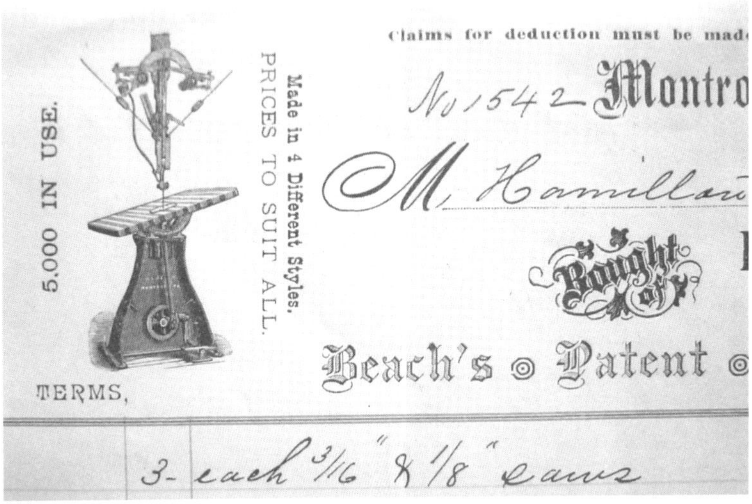

Receipt showing a detail of the first scrollsaws Ed Hamilton used to make Holly Wood Type.

Chapter 2

THE GROWTH OF HAMILTON

As the United States emerged from the economic depression that followed the Civil War, opportunities for new businesses began to develop in the rapidly expanding midwestern states. Due to the increasing distances between manufacturing facilities and the consumers in need of products, the centralization of industrial production in the Northeast was no longer as efficient as it had been before the War. Also, as the economy improved and the frontier towns in the West grew, the need for low-cost, semi-durable goods like wood type and other printing supplies offered a multitude of opportunities for men of drive and business acumen.

A rendering of the Hamilton Katz Holly Wood Type Company factory from the late 1880's.

Receipts from the early days of the Hamilton Manufacturing Company.

It was into this rich economic environment that Ed Hamilton's wood type manufacturing business was born.

In the 1880s, when Hamilton invented the process for manufacturing Holly Wood Type, all wood type was being manufactured in the East. But as newspapers began popping up in towns west of the Mississippi and the need for type grew, the time and expense required to get good, usable type from the manufacturers in the East became a serious problem for printers. Holly Wood Type was the perfect solution to the problem.

Possessing only limited, first-hand experience in running a business, Ed and Etta Hamilton did their best to keep up with the rising demand for their type, recording orders on the wall alongside the bench in the shop attached to the back of Ed's mother's house. The shop doubled as a nursery when their first child, Grace, was born. They soon realized that such an improvised system of tracking and delivering orders was not going to work, so they bought and studied a book on bookkeeping. In later years, Ed recalled he and Etta had no idea how to ship their type to their customers. The Hamiltons ended up relying on the assistance of the local railroad and express agent, who showed Ed how to fill out the necessary papers to ship and receive payment for his type.

Ed Hamilton made frequent trips from Two Rivers to Milwaukee in search of new type orders and ideas that would appeal to his printing customers. On one such trip Ed met with Henry Katz, a friend of the principal owners of the chair and pail factory where Ed had once worked. Katz was intrigued by the plans and business prospects of the young man with a case full of wood type samples. After checking with some of his contacts in Two Rivers, Katz suggested to the young type-maker that he might benefit from selling a portion of his business to raise capital. Although leery of taking on a partner, Ed recognized that he needed money to expand, and after some negotiation, sold half the interest in J.E. Hamilton Hollywood Type Co. to Katz's son Max.

The money raised by the sale of a share of his business to Max Katz allowed Hamilton to build his first factory.

According to Two Rivers' legend, the move into the new factory - actually a barn with an office built into the corner of the lower floor - was accomplished using a goat wagon borrowed from one of Ed's neighbors. A 'hands-on' manager for his entire career, Ed walked alongside the cart, steadying the top-heavy load of stencils, woodcutting tools, foot-powered scroll saw and office supplies.

Expanded Hamilton Manufacturing Company plant at the turn of the century.

Katz retired from the business after a few years, selling his interest to William Baker, a part owner of a defunct sash and door company in Two Rivers. Ed had been interested in the empty sash and door factory for some time, and used his connections with Baker to buy the plant. The old sash and door factory remains part of the Fisher Hamilton facility to this day.

Demand for wood type and related materials in the United States continued to grow and its increased production capacity allowed Hamilton to grow along with it. World events combined with the astounding growth of the newspaper industry to

*A Hamilton Manufacturing Company ad as it appeared in the **Inland Printer** magazine near the turn of the century.*

increase the need for printing materials in the American West. Ed Hamilton's initiative and business acumen put his company in a good position to take advantage opportunities offered by the expanding national economy.

With the approach of the 20th century, the United States was beginning to be an important actor on the world stage. Fueled by the economic benefits of westward expansion, the United States was turning outward, challenging the hold of traditional imperial powers on the markets in the colonies in the Southern Hemisphere. As the relatively young country came to terms with its new role in the international community, newspapers also began to turn their attention to international issues, covering - often sensationally - the goings-on in such far flung places as the Philippines, Africa, and the Far East.

The growth of cities in the American West led to a boom in newspaper publishing, and with it, an increased demand for attention-grabbing headlines designed to sell papers. This sensational approach to newspaper reporting, which eventually became known as "yellow journalism," appealed to a jingoistic streak in the American public and focused newspaper readership on international issues. American industrialists worked with newspaper publishers to put pressure on the government to be more active in opening traditionally colonial markets, feeding the newspapers stories of real and imagined atrocities perpetrated by colonial powers to inflame the voters.

The most obvious example of the use of "yellow journalism" to drive actions on the international stage was the publication sensationalist stories and dubious headlines to end Spain's hold on a number

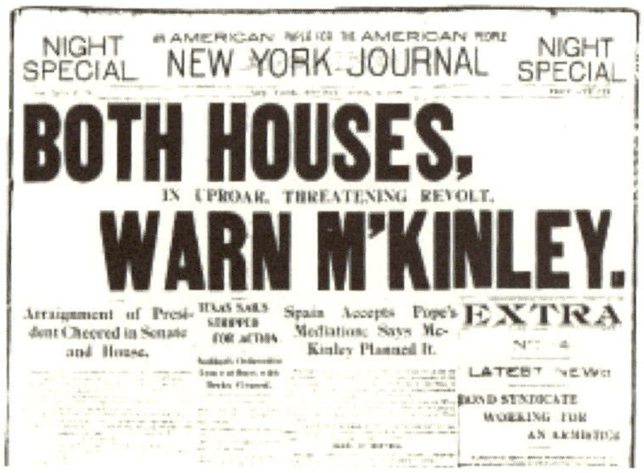

NIGHT SPECIAL NEW YORK JOURNAL NIGHT SPECIAL

AN AMERICAN PAPER FOR THE AMERICAN PEOPLE

BOTH HOUSES,

IN UPROAR, THREATENING REVOLT,

WARN M'KINLEY.

Arraignment of President Cheered in Senate and House.

TEXAS SAILS STRIPPED FOR ACTION

Spain Accepts Pope's Mediation; Says McKinley Planned It.

EXTRA

LATEST NEWS

BOND SYNDICATE WORKING FOR AN ARMISTICE

Sensational newspaper headlines of this type led President McKinley to declare war on Spain.

of its colonies. Stories of Spain's actions governing its colonies in Cuba, Puerto Rico and the Philippines were given prominent, if not always accurate, coverage in American papers. The Hearst and Pulitzer newspaper chains were locked in intense competition for readership and used exaggerated accounts and spectacular illustrations to sell newspapers. The pressure on President McKinley to save Spain's colonies from its oppressive rule as reported in the nation's newspapers is often credited with leading the United States into the Spanish-American War in 1898.

The Hamilton Company, as it had become known, was well positioned to take advantage of America's increasing appetite for newspapers. While many of the dailies in the larger cities had moved to linotype for page layout, papers in smaller cities, both dailies and weeklies, still did much of their assembly by hand using movable type. Further, many of these smaller papers augmented their income by producing handbills and posters for local businesses. These side businesses regularly required reasonably-priced, large-scale type. The Hamilton Company had the production facilities, sales contacts and distribution network in place to meet the demand.

Poster showing a typical diversity of wood type.

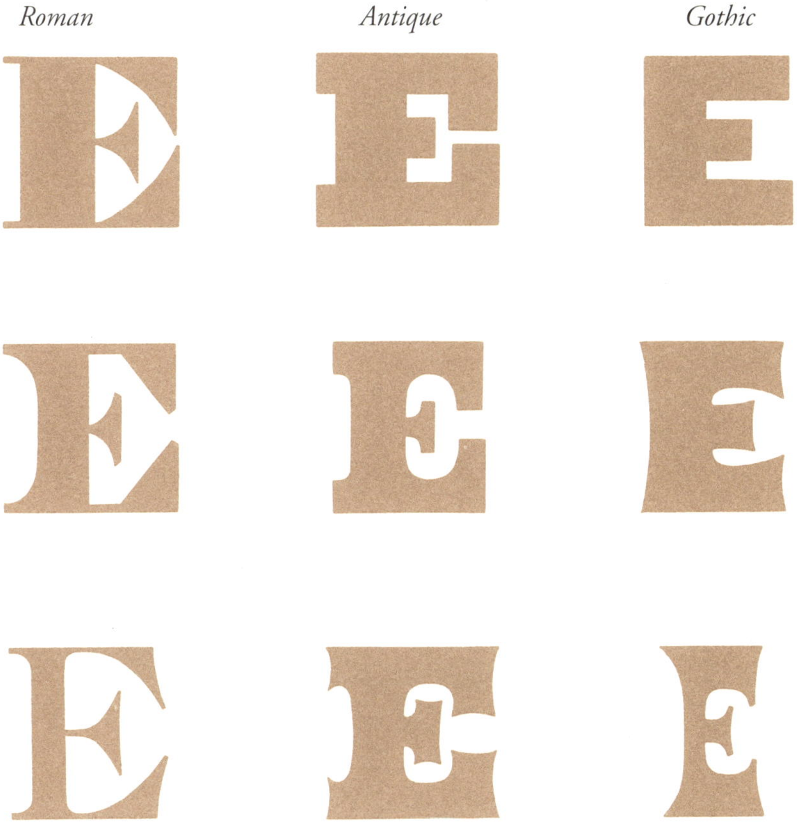

The grid above shows the three primary families of wood typefaces. Roman, Antique and Gothic with variations shown in the columns below.

Eventually the company abandoned the veneer-style of manufacturing type for the more traditional and durable end-grain manufacturing approach, bringing it into direct competition with the major wood type companies in the East. With a plentiful supply of Wisconsin rock maple and Ed Hamilton's gift for understanding and improving industrial processes, the Hamilton Company soon became one of the most important suppliers of wood type in the world.

Although Ed Hamilton had little business experience when he started cutting out type with his scroll saw, he learned the type business quickly and began to find new ways to build his company's share of the wood type market. In addition to making high quality type more cheaply than other companies, Hamilton also came to understand that providing printers with an easy way to order a wide variety of font styles was another way to expand the business.

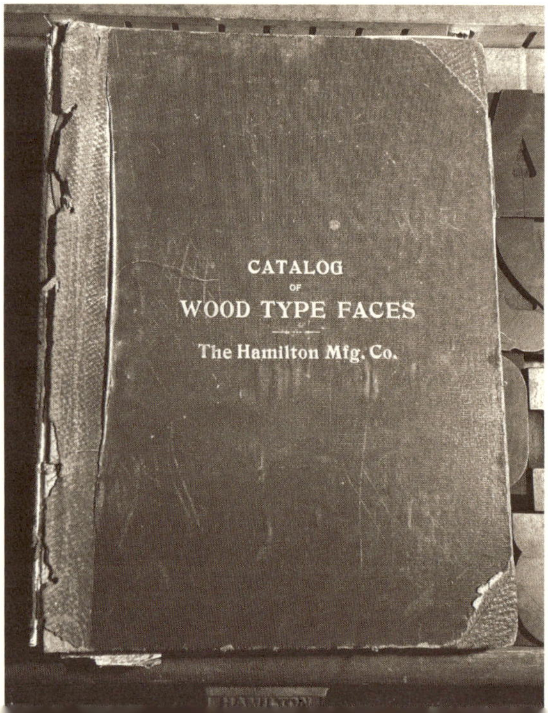

It was common practice for wood type companies to "borrow" the designs of their competitors and recut their own versions. Antique, Roman and Gothic are examples of popular faces of the time. During the 1890s, Hamilton began acquiring other wood type companies and found himself having to manage an extensive catalog of designs while supplying printers with a consistent product. If a printer wanted Wells version of Egyptian Extended, it was Hamilton's obligation to send the correct face. To that end, Hamilton Manufacturing released a now rare comparative specimen book in 1906, allowing salesman to correctly identify which version of each face belonged to whom.

Understanding the importance of maintaining control over its intellectual property, Hamilton included a special notice in its specimen book:

No. 5056—Class O Heber Wells' Egyptian No. 2 6 Line, 6c per Letter

REN dan 26

No. 4160—Class N Wm. H. Page Co.'s Belgian 6 Line, 6c per Letter

RENACDKMSUV damecnrsuvh 23456

No. 3055—Class N Morgan & Wilcox Mfg. Co.'s Belgian 6 Line, 6c per Letter

RENACDGHMSZ damersuwh 23456

No. 160—Class N Hamilton Mfg. Co.'s Belgian 6 Line, 6c per Letter

RENACDGSTUZ damebcrsuh 234576

No. 5057—Class N Heber Wells' Belgian 6 Line, 6c per Letter

RENABDGHUS damcrsuh 234567

No. 191—Class N Hamilton Mfg. Co.'s Belgian No. 1 6 Line, 6c per Letter

RENACDSTU damecrsh 234567

No. 5058—Class N Heber Wells' Belgian No. 1 6 Line, 6c per Letter

RENACDGHJLMOSTUWZ

No. 279—Class N Wm. H. Page Co.'s No. 155 6 Line, 6c per Letter

RENACDGST damersh 2456

All Type shown can be made in any size desired.

Cover and sample page from one of Hamilton's first comprehensive specimen books, circa 1906.

This Wood Type Catalog No. 16, issued by the HAMILTON MANUFACTURING COMPANY, is not published for free distribution. The edition is limited and numbered, and whenever these books leave our hands, it is with the distinct understanding that the book is to remain our property, and subject to recall at any time. Under no other conditions will one of these books be sent to an inquiring customer. It has taken two generations to accumulate all of the material shown in this book, and for this reason we must protect our interests. Something like 1,500 different Designs are shown in this specimen book and it comprises the largest assortment of Type designs ever shown under one cover.

Hamilton's No. 1 Double Polhemnus Cabinet designed to store and compose type.

The Hamilton specimen book is the only known resource showing comparative font faces of all the major American manufacturers at the height of wood type's popularity.

In addition to "borrowing" designs from other type manufacturers, Hamilton extended its control of the wood type market by buying established type companies. The acquired company's unique fonts were folded into the Hamilton line, and Hamilton's ability to mass produce high-quality type allowed it so sell the product at lower prices. By 1900, the Hamilton Company had a virtual monopoly on the production of wood type in the United States and had begun shipping its type to printers across the world.

As his wood type business grew, Ed Hamilton spent a lot of time in print shops, which he found uniformly dirty and disorganized. Looking to expand his line of printer's supplies, Hamilton saw a business opportunity in the design and manufacture of furniture and other supplies designed specifically for use in print shops. The equipment the company acquired when it bought the old sash and door factory was well suited for the manufacture of wood furniture, and the company launched a new line of printer's supplies.

The printer's cabinetry side of the business grew rapidly, and by 1910 it represented a full two-thirds of Hamilton's business. The company soon added a similar line of furniture for dental offices. Dentists needed storage for tools and devices in their offices, and Hamilton's experience meeting the needs of print shops transferred well to the design and manufacture of this specialized furniture.

Encouraged by the success of its dental office line, Hamilton developed and launched a line of drafting room equipment in 1917. Within ten years, the Hamilton Company became the principal manufacturer

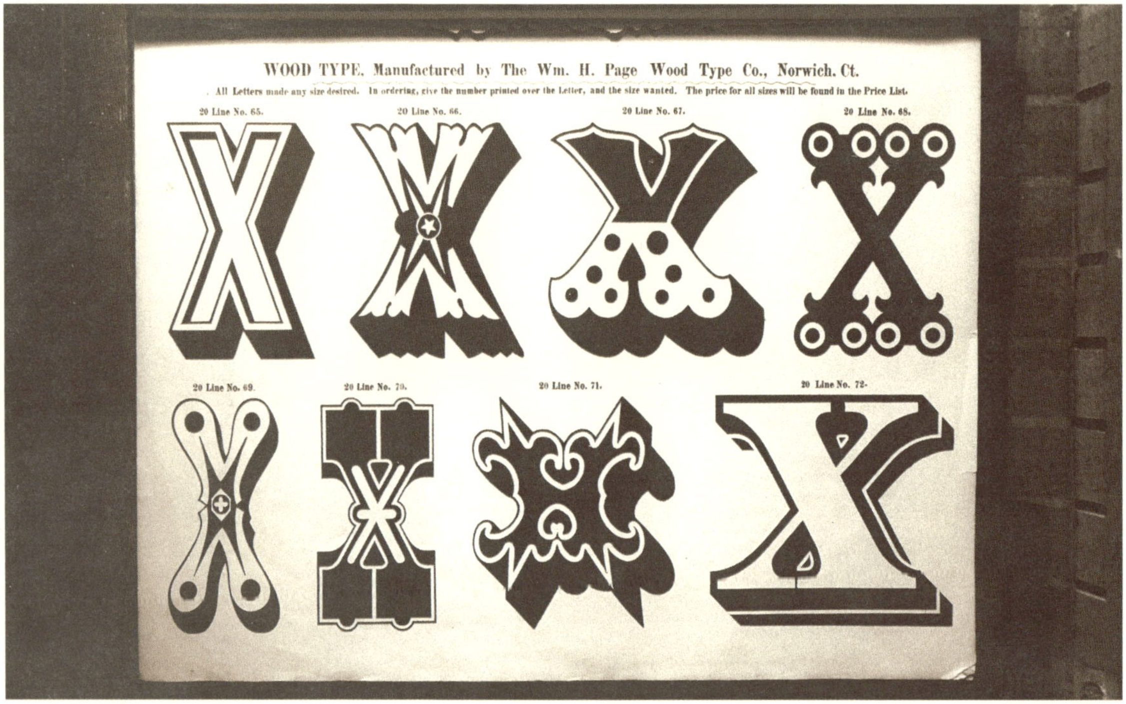

Specimen sheet from the William H. Page Company showing the ornamentation in wood typefaces.

of drafting room storage systems and furniture in the United States. The company also used its woodworking facilities to make attractive cabinets for the rapidly expanding console radio market.

Around that time, the Hamilton Company began experimenting with using steel in the manufacture of its printers' furniture rather than wood. Not satisfied by the equipment turned out by "experts" brought in to advise the company on working with metal, Hamilton brought the development of metal printers' furniture in-house, where the needs of their specialized market were better understood. The sale of this steel furniture soon outpaced the traditional line of wood printers' furniture.

In later years, the company's expertise in making specialized printer's furniture with steel allowed it to move into the manufacture of other metal products, including laboratory equipment, doctor's office furniture, and the country's first automatic clothes dryer.

By the time J.E. Hamilton retired in 1919, he had established a culture of quality, pride and innovation in the company that today still bears his name. Although he and his family stayed involved in the community of Two Rivers through their generous support of a variety of charitable causes, his lasting legacy resulted from from his ability to recognize and capitalize on the business opportunities he found in the rapidly expanding economy of the United States at the turn of the century.

That, and his ability to cut wood type.

Ed Hamilton had a keen sense of the importance of marketing. His company's name and signature craftsmanship could be found on every surface including drawer pulls.

Chapter 3
THE MAKING OF WOOD TYPE

Wood has been used to transfer ink onto a printed surface for centuries. Chinese printers were known to use wood in block printing as early as 770 A.D.. In Europe, religious books made using carved wood blocks appeared many years before the invention of movable metal type. As the complexity and ornamentation of printed material grew, wood block illustrations, decorative borders and large type figures became important tools in every printmakers' shop.

This type of router, (above right) featuring a movable bit, was employed by the E.R. Webb Company in the 1850s.

The ability of wood type to produce clean, consistent characters in larger type faces made it well-suited to producing larger printed pieces such as posters and broadsheets. Large metal types were expensive and difficult to handle, often failing to provide even ink coverage because their printing surfaces were somewhat concave due to the uneven cooling of metal. Wood type, on the other hand, could be made to provide large, flat printing surfaces, making it perfect for larger-scale letter forms.

At the beginning of the nineteenth century, wood type was produced largely by hand. Most wood type was not used directly in the printing process, but rather as dies for sand molds used to make metal type. The growth of the commercial printing industry in the United States in the early part of the century along with two inventions introduced by American printers allowed the use of wood type to become an important part of the printing business.

Half round slab of rock maple, the raw material from which Hamilton wood type was made.

Although various attempts were made to automate the process, none met with the success until New York-based printer Darius Wells' introduced the lateral router in 1827. Using a low-speed router suspended from the ceiling of his plant, Wells was able to produce many more finished pieces of type in a week than had been possible using traditional methods. Wells' use of the lateral router was so successful that he was able to produce the first known wood type catalog in March of 1828.

One of the main problems with hand-carved wood type was the difficulty producing consistent product from one set to another. In 1834, George Leavenworth added a pantograph (a device used to trace figures at different scales onto surfaces) to Wells' lateral router, making it possible to use a single set of patterns to produce type of different sizes, assuring product consistency from one set of letters to another.

Holly Wood Type

Ed Hamilton's chance introduction to the wood type business resulted more from his woodworking skill and entrepreneurial spirit than any interest in the printing business.

Most wood type used after the Civil War was the end-cut variety, carved using the combined pantograph and lateral router. But when William Nash asked Ed Hamilton to make him some type, regular wood type making equipment was unavailable. A different approach was needed. Rather than routing the design out of the end grain of a block of hardwood, Hamilton traced his design on a thin piece of holly wood, selected because it works easily without splitting and has a clear, close grain that allows it to be polished to a good printing surface. He then used his foot-powered scroll saw to cut the letters, incised into the wood so they would print white on a black background. The design was mounted on a softer piece of wood, sanded and polished to a printing-quality finish. It was then easy to plane the underside of the block until it was approximately "type-high."

Hamilton's veneer approach to producing type was a significant, albeit temporary, step forward in the production and distribution of wood type in the United States. Produced from readily available materials and using fairly standard woodworking tools, Holly Wood Type could be sold for roughly half the price of the end-cut wood types available at the time. And although the veneer type was not quite as durable as end-cut types, it held up well enough to allow printers of moderate means to compete with larger operations for poster and other large scale printing projects.

By the time Hamilton Company stopped making Holly Wood Type in favor of the more traditional end-cut type sometime around 1890, it had established itself as one of the prime forces in the wood type business. The company's ability to compete in the type market was the direct result of Ed Hamilton's success with Holly Wood Type.

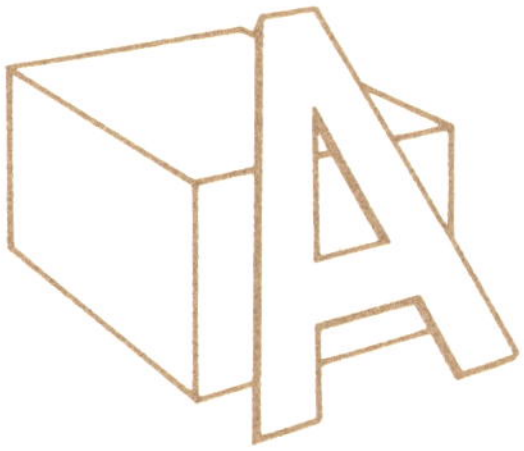

Holly Wood Type as invented by Hamilton is produced by laminating a thin veneer of cut wood on top of a larger block.

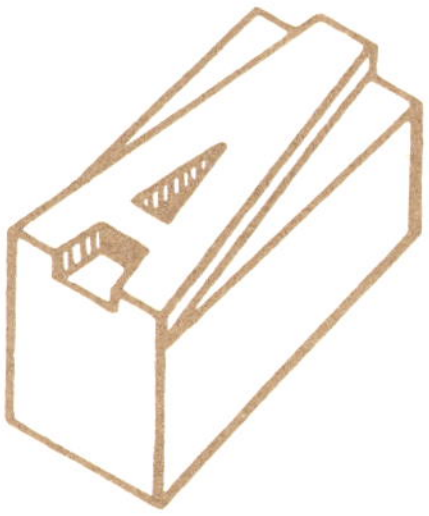

End-grain wood type is routed from a single block of wood.

Maple slabs in drying racks before the first round of surface finishing.

End-Cut Wood Type

Although many manufacturers experimented with different methods for mass-producing wood type, the end-cut method proved to be the best and most reliable way of producing clear, consistent typefaces out of wood at a reasonable cost.

End-cut wood type is made in a process similar to one favored by wood engravers for generations. Basically, the design is transferred onto the end grain of a block of wood, carved, then polished to yield a printable image. By employing the pantograph, manufacturers were able to transfer consistent designs of varying sizes to the wood blocks and the lateral router made it easy to quickly remove the wood and produce the printing block. Although some wood type manufacturers near the turn of the century cut type designs into the side grain of the wood (notably George Leavenworth), end-cut characters tended to print a little better and lasted longer so were generally preferred by printers.

The preparation of the wood is a key element in the production of good end-cut type. It is also the most time-consuming part of the process. At the height of its wood type production, Hamilton Company had five-foot lengths of split maple logs delivered to the plant by rail sometime in February. The logs were then cut into 18-inch lengths to allow workers to easily handle them. The 18-inch sections were cut into one-and-a-quarter-inch slabs so they could be placed in the drying kiln. The bark was left on the slabs, allowing the wood to dry slowly, minimizing splitting and warping.

Museum volunteer and former Hamilton employee Gene Petrie cutting maple logs into slabs.

Dried maple slabs waiting to be sanded to type high.

Museum volunteer and former Hamilton employee Norbert Brylski cuts a maple slab to size on the line saw.

The one-and-a-quarter-inch slabs were placed on racks in the drying kiln where the heat, moisture content and air circulation around them was tightly controlled. The goal of the drying process was to to reduce the moisture content of the slabs from nearly 100% to around 18% so they could be piled on top of one another to conserve space. Drying the slabs in the kiln took from six months to a year, after which the slabs were removed from the kiln and placed in a tempering shed to equalize the moisture in each of the slabs. A band saw was then used to remove the bark from the slab, providing a clean, dirt free block of wood.

The printing surface of the wood was then prepared. The top side of the cured slabs were machine-sanded to a high polish and the bottom was planed to "type-high" thickness, approximately 0.918 inches. Hamilton quality standards allowed no more than 0.003 inches of variation throughout each slab. The top side of the slab was sanded again, this time by hand, and sprayed with a sealant. Once dry, the slabs were hand polished to provide the final, printing-ready surface.

With the surfaces of the wood slabs prepared and their thickness made uniform, the slabs were then cut into the blocks. First the heart-wood and other unusable sections were removed, then the slab was cut down from the largest piece to the smallest. Carefully designed cutting templates were used to make sure the most was made of each slab.

Museum volunteer Mardelle Doubek, one of the last Hamilton type cutters, using a pantograph to cut a type block.

End-grain type block with an inset showing how two different routers and hand tools were used to cut the type.

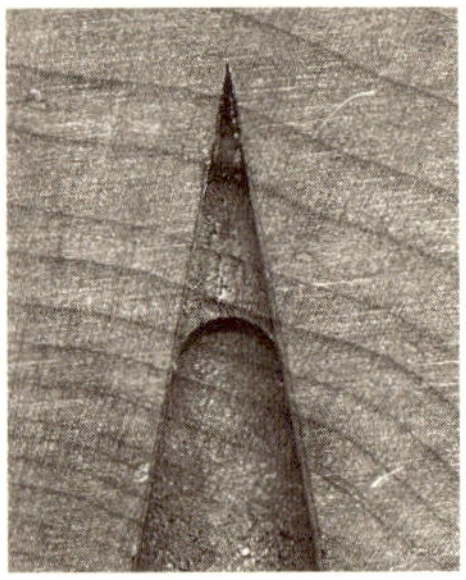

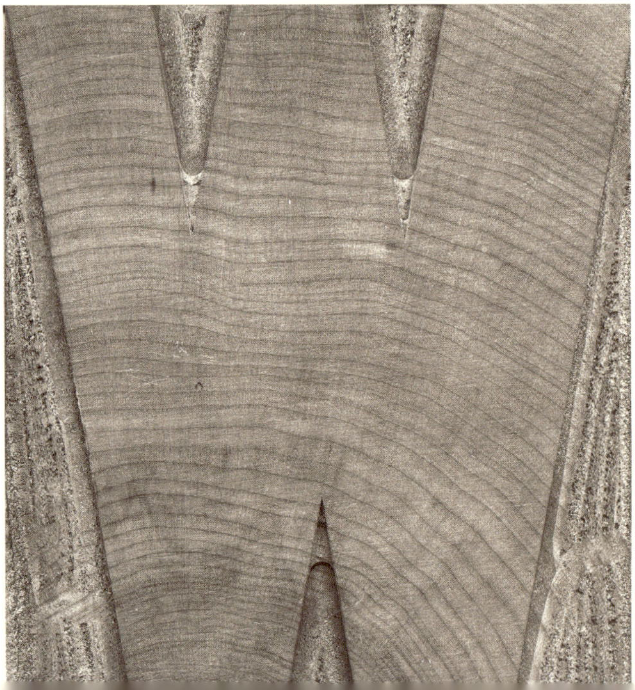

The sized blocks were then ready to have characters routed onto them. Workers used a pantograph to trace the pattern of the letter while the attached lateral router reproduced the pattern on the wood according to the size of type required. An advantage of the pantograph was a single set of type face patterns could be used to make type sets of any size. The routing of each block was a two-step process: first a large router bit was used to remove most of the wood from the block, then a fine router bit was used to further refine the interior of the letter forms.

The blocks were then trimmed down to provide the proper printer's spacing and alignment.

Finally, the finished type blocks were taken to type trimmers who used hand tools to put the final touches on each piece. This was an intricate job requiring a watchmaker's precision and attention to detail. Each trimmer not only had to know which corners of the letter forms were to be pointed and which rounded, but they were also responsible for inspecting the type as a final quality check. The finished type pieces were then assembled into sets on the packing bench, inspected a final time, then wrapped for shipment to print shops in the United States and around the world.

Hamilton retiree Bernice Schwahert demonstrates her skill as a type trimmer.

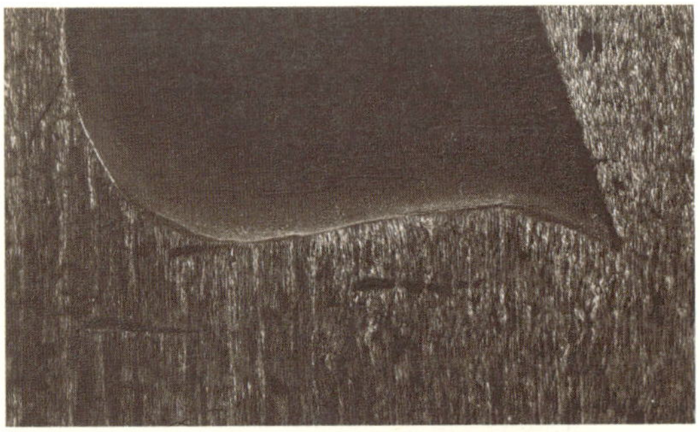

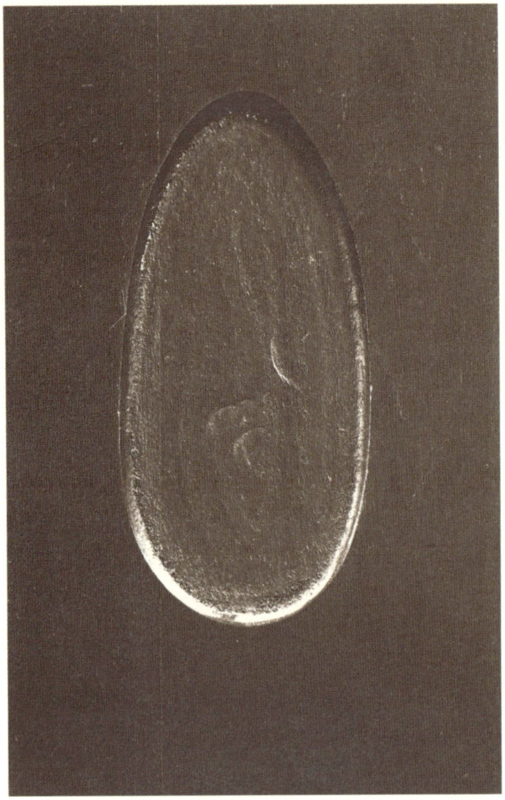

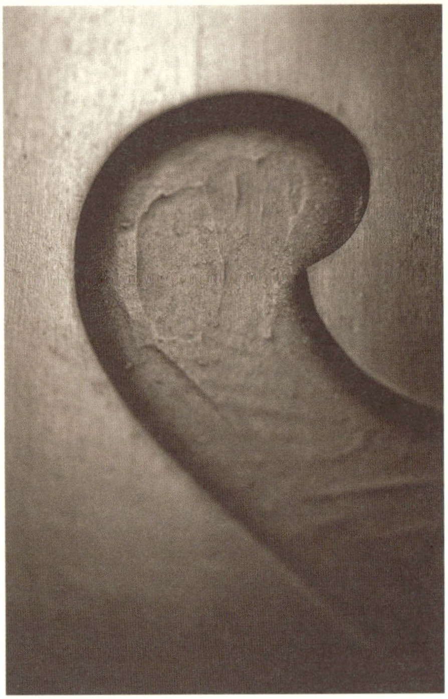

These examples show the detailed interiors of letter forms. Wood types signature look is defined by the irregular counters (non-printing area). Standard letterforms had to be modified to yield as much printing area as possible, in order to achieve the amplified effect for which wood type is known.

Hamilton's Grecian Extra Condensed.

Chapter 4
HAMILTON REOPENS ITS DOORS

May 28, 1999 dawned bright on the western shore of Lake Michigan. The City of Two Rivers was gearing up for summer, when tourists with fishing poles and time on their hands descend on the city to spend their vacation days drowning worms, shopping for antiques, barbequing bratwurst and enjoying the cool breeze off the Big Lake.

Just a bit inland, some of Two Rivers' citizens had come together to get their first look at a new attraction, the culmination of five years of planning, good-luck, creative fund-raising and old-fashioned sweat-of-the-brow hard work on the part of a dedicated group of local preservationists. They had come to visit a new museum that both honors the heritage of an industry that meant so much to Two Rivers and provides a destination for artists and designers from around the world interested in an almost forgotten printing tool. The Two Rivers Historical Society was proudly introducing the most comprehensive collection of wood type tools and artifacts in the US: the Hamilton Wood Type and Printing Museum.

The Museum on opening day.

Standing before an audience of printers, former Hamilton Wood Type employees, design professionals, printing devotees and Two Rivers residents, then-Wisconsin Governor Tommy Thompson commended the community for their hard work and determination in preserving this important piece of the state's history. Joining Thompson in welcoming the new museum were Two Rivers city manager Greg Buckly and J. Michael Brown, president of Fisher Hamilton, Museum sponsor and head of the company that traces it roots directly back to the wood type business. Also on the flatbed trailer that doubled as a receiving stand were executives from Fisher Hamilton, local politicians and select members of the Two Rivers Historical Society. At the center of it all was James Van Lanen, Sr., Historical Society member and local businessman without whose vision and persistence the Museum would never have become a reality.

Jim Van Lanen, Sr.

After the official festivities ended, many in attendance went across the street to savor another local attraction – a Breners Ice Cream Sundae. Others strolled under the Museum's drawer-pull shaped front entrance, stepping into a nearly forgotten slice of Two Rivers' history.

Wisconsin Governor Tommy Thompson at the Hamilton Wood Type and Printing Museum opening.

The Collection

On opening day, the Hamilton Wood Type Museum occupied 30,000 square feet of space on the site of Fisher Hamilton, very near the original Hamilton Manufacturing Company factory. Most of the Museum space was dedicated to tools and artifacts related to the process of wood type manufacture, with approximately 4,500 square feet set aside for a printing facility to demonstrate how Hamilton wood type was used to produce broadsides, newspaper headlines and posters.

Pantographs used to cut type.

In the wood type manufacturing area, the Museum had on display all the equipment required to convert a maple log into finished wood type. Nearly all the tools and machinery in the collection were made in the Hamilton Wood Type Company plant and, since much of this equipment was unique and specialized, the Museum relied on former Hamilton workers to help them assemble and learn how to use the equipment. All the saws, routers, pantographs and patterns needed to make wood type are in full working order.

The equipment was used regularly by volunteers, many of whom are former Hamilton workers, to demonstrate how type was made. Demonstrations of the production of wood type were made on request and when the Museum was producing commissioned type sets.

The printing facility featured three Vandercook proof presses, drying racks, type cases, composing tables, paper cutters, linotype and intertype machines - a complete print shop used by master printers and their students to learn from the unique properties of wood type. Most of the equipment in the shop was donated by a host of printing firms as well as individual printers.

One of the more unique items on display in the print shop is a ruling machine. Until 1966 the machine was used by Bernie Moran, a second-generation printer from Green Bay, WI, to print ledger pages for accounting books.

Toto in Oz by Richard Zauft.

A Morgan sign press with 3,250 pieces of metal type and 3,500 pieces of wood type was also on display when the Museum opened.

The printing facility has been used by a number of nationally recognized master printers, including Richard Zauft, Dennis Ichiyama, Ruth McGurk and Mark Snyder. Zauft and Ichiyama have also brought advanced design and printmaking students to work with wood type in the museum' print shop, introducing the aesthetic possibilities of this nearly-forgotten medium to a whole new generation of printers.

Visitors outside the Museum on opening day.

In addition to the manufacturing and printing operations, the Hamilton Wood Type Museum featured at the time of its opening:

- more than 1.5 million pieces of wood type,
- more than 1,000 different styles and sizes of type patterns,
- a wall of wood poster type eight feet high and 145 feet long, the largest such collection on public display in the US,
- type cut by Darius Wells, father of the US wood type industry, dating back to 1840,
- type ranging in size from 3 line (1/4 inch) to a pattern for a number '2' that measures 56 inches high by 72 inches wide,
- an extensive collection of broadsides, newspapers and posters printed using Hamilton type,
- many of the Hamilton type catalogs, and
- photographs and other memorabilia documenting the history of Hamilton Wood Type Company and the life of its founder, James Edward Hamilton.

Image made using one of the oldest examples of Hamilton type on display at the Museum.

One of the oldest examples of Hamilton type on display is a set of three engraved wood blocks used to print a three-color image of an ear of corn. The blocks, made sometime between 1892 and 1894, were donated to the Museum by Cliff Harvey of West Virginia University.

The Museum's collection also includes complete sets of two new type faces commissioned specifically to be rendered in wood. One of these new font styles, Carter Latin, was designed for the Museum by world-famous British type designer Mathew Carter. The other, Hamilton Offset, was developed by the author Bill Moran.

A section of the largest display of wood type in the United States.

History of the Museum

During the 1990's, the Two Rivers Historical Society became involved in a series of preservation projects that ultimately led the to launch of the Hamilton Wood Type Museum. The Historical Society, seeking to develop tourist destinations in Two Rivers that celebrate the town's rich history, had begun acquiring properties of significance to prevent their destruction.

The timber and maritime industries have played a important role in the development of the Two Rivers area. So it is fitting that these two cornerstones of the community would, coincidentally, come together to give birth to Hamilton Wood Type Museum.

In the early 1990s, the City of Two Rivers funded a study to determine feasibility of opening a Coast Guard Museum in town. Jim Van Lanen Sr.'s experience owning and operating the Lighthouse Inn made him quite familiar with the tourism trade in the region. He was a natural choice for the city when they were looking for a local preservationist to take the lead on assessing this proposed attraction. Once the city decided to go ahead with the project, Van Lanen was chosen to see the museum to completion. The experience he gained preparing the museum space, assembling the collection, securing resources and operating the museum for the first two years of its existence proved invaluable in his later work on the Hamilton Museum.

Wood type on display at the Museum.

The Fisher Hamilton Company stopped making wood type in 1985, selling the business, equipment and patterns to Jim Kearns, their largest wood type customer at that time. Fisher Hamilton moved on from its roots, concentrating on its core businesses of laboratory furniture and air flow systems while Kearns continued to use the equipment to produce type until 1993.

In the fall of 1993, Van Lanen received a call from Gib St. Pierre, a local electrical contractor. Mr. St. Pierre had been told that someone from a museum in Texas was coming to town to take a look at the old Hamilton Wood Type equipment. Concerned that this important part of Two Rivers history was about to be sold and moved out of state, Van Lanen immediately went to Kearns' place of business, introduced himself, and inquired about the old Hamilton equipment, stock and patterns. Kearns took Van Lanen on a tour of the type-making equipment and agreed to meet with the Two Rivers Historical Society in the near future to discuss keeping the collection in Two Rivers.

After meeting with the Historical Society and providing another tour of the equipment, Kearns rejected Van Lanen's proposal that he donate the machinery and stock to a wood type museum operated by the Society. In negotiations with Van Lanen, Kearns offered to sell the equipment to the Society at a below-market price and Van Lanen began looking for a space that could house a wood type museum.

After securing a low-cost space in a building owned by the city, Van Lanen presented a proposal to the Historical Society to purchase the old Hamilton equipment and set up a museum in the currently vacant space. The proposal met with some resistance from most of the Society. Many board members felt that this kind of project was too ambitious for the group and the idea was never brought to a vote. The idea of a wood type museum was officially dropped.

Van Lanen did not give up. He maintained contact with Kearns through the early part of 1994, as well as placing calls to print museums in Texas and California to further research the value of the old Hamilton equipment and the historical importance Hamilton Wood Type. As Van Lanen learned more about the role Hamilton played in the development of the printing industry by providing inexpensive, high quality wood type to the rapidly expanding market, he became convinced that he needed to do all he could to preserve this essential piece of Two Rivers history. But the project needed more muscle - there was no way the Historical Society could do this on its own.

During a tour of the Fisher Hamilton offices in the Fall of 1994, Jim Van Lanen shared his idea for a museum of wood type to Dave Pflieger, an executive with the company. Pflieger was intrigued and agreed to set up a meeting between Van Lanen and then company president Carl Bretko to discuss getting the project off the ground. Two weeks later, the three had lunch and the museum project had new life.

Bretko loved the idea of a Hamilton Museum in Two Rivers. And recognizing the importance of keeping the old equipment in town, he volunteered to take a look at buildings on the Fisher Hamilton site to see if a space could be found to house the collection. With the support of the company president, Fisher Hamilton quickly identified a suitable space on their campus and discussions began in earnest between the company and the Historical Society on making the Hamilton Wood Type Museum a reality.

Van Lanen reopened discussions with Jim Kearns regarding the purchase of the old Hamilton Wood Type equipment. After some back-and-forth, Kearns agreed to sell the majority of the equipment and supplies he purchased with the Hamilton Wood Type name to the Historical Society. The Historical Society worked closely with a local bank to secure financing for the purchase.

Fisher Hamilton prepared to receive the collection by clearing the Museum space and repairing a leaky roof. Van Lanen and other members of the Historical Society spent weeks disconnecting and shrink-wrapping equipment, packing up catalogs and preparing skids of wood to be moved. By the time they were finished, the volunteers and Fisher Hamilton employees assembled and moved thirteen semi and 35 pickup truck loads of equipment. The move took eight men 10 days of work to complete.

The purchase of the Hamilton Wood Type equipment was finalized in February of 1995 and by the end of March the saws, patterns, pantographs, lateral routers and wood stock were on the Fisher Hamilton site. The company generously agreed to lease and provide utilities for nearly 30,000 square feet of space for the Museum, asking only for $1 a year in rent from the Two Rivers Historical Society.

Recognizing the need for experienced input in the development of the space, the Historical Society hired The Design Office of Catherine and James Donnelly of Milwaukee to put together a floor plan and design artifact display areas for the new Museum. James Donnelly in particular became very involved in the project, making many trips to Two Rivers to oversee the implementation of their design.

Donnelly, in turn, solicited input from Richard Zauft, a letterset printing expert and Associate Dean of the Peck School of the Arts at the University of Wisconsin-Milwaukee. Zauft would become another important player in the Museum project, assisting in the development of the on-site printing operation at the Museum and promoting the museum through his contacts in the worlds of printing and design.

A museum volunteer sorts patterns in the new museum.

During the four years it took to bring the museum together, the Society assembled a dedicated group of volunteers without whom the project could not have been completed. This collection of local preservationists, printing enthusiasts and current and former Hamilton employees set up equipment, painted walls, made display cases and just about anything else that was needed to get the building ready to receive visitors. The Hamilton retirees club was very helpful in showing how the various pieces of type making equipment were set up and used. The involvement of many of these volunteers continues to this day and provides an important link to the people who produced the wood type that made the country's headlines for three-quarters of a century.

In order to provide financial support for the Museum, Van Lanen and the Historical Society organized a fund drive in 1997 that resulted in a grant of $50,000 from the Ruth St. John and John Dunham West Foundation. The fund drive also benefitted from a partnership with Oak Knoll Books, publisher of "books about books," that included information on the Museum in its regular mailings. Many of these solicitations resulted in donations from printers and other design professionals from across the country.

As the museum project moved near completion, Van Lanen and the Two Rivers Historical Society realized they had a bit of a problem: they had the equipment and raw materials to make wood type, but very few examples of the finished product.

This lack of type needed to be addressed. After all, a wood type museum without wood type was like an ice cream sundae without the ice cream.

Enter Irving and Nancy Silverman. The Silvermans had been avid collectors of wood type for forty years, amassing a huge collection of Hamilton-produced type in a wide variety of styles and sizes. Their collecting days were nearing an end, and Mr. and Mrs. Silverman were looking for a good home for their collection. Irving assisted the Two Rivers Historical Society in raising $40,000 to purchase the collection as well as generously donating another $80,000 worth of type. The contribution made by the Silvermans comprised the majority of wood type in the Museum's opening day collection.

Shortly after coming into possession of the Silverman collection, the Museum was approached by Leo Kaplan and his wife with an interesting proposal. The Kaplans were also wood type collectors and had 15 cabinets full of type they would donate if the Museum would cover the cost of the move. David DeRosier, the owner of Wisconsin Nation Wide Trucking Company and a friend of the Two Rivers Historical Society, graciously arranged to have the type cabinets picked up and delivered to the Museum free-of-charge.

The Hamilton Wood Type Museum stands as a monument to James Edward Hamilton, his family, Hamilton Company employees and all the Two Rivers residents who helped build the printing industry in America. And through the efforts of the many printers, designers and students who travel to Two Rivers from all over the world, the legacy and distinctive aesthetic of wood type printing will be preserved for generations to come.

A museum volunteer moves type cabinets into their final position.

Chapter 5
THE LIVING MUSEUM

As our world becomes increasingly digitized, preserving the techniques used to manufacture and print with wood type becomes even more important. Although laser printers, digital imagery and direct to press production have become staples of the commercial printing industry, interest in the traditional letterpress printing remains strong. More than a regionally significant curiosity, the collection at the Hamilton Wood Type and Printing Museum keeps alive the unique properties of wood type for printers across the world. In the essays that follow two nationally-recognized master printers and educators describe the ways in which they have used the Museum and its collection to bring the possibilities of wood type to a new generation of designers and print artists.

Contemporary Uses of Wood Type
by Richard Zauft

Richard Zauft is a professor and associate dean at the Peck School of Arts at the University of Wisconsin-Milwaukee where he teaches letterpress printing, the history of design and typography. He has been a friend and supporter of the Museum since its inception, promoting the collection in lectures and workshops.

I became involved with the future Museum in 1997 when James Donnelly, a designer in Milwaukee who was initially hired to design the Museum interior, contacted me because he knew I was a letterpress printer. Donnelly and his staff were designing the floor layout, the displays and Museum signage, so he invited me to drive up to Two Rivers with him for one of their planning meetings so I could see what was going on. The opportunity to see thousands of pieces of wood type, and the possibility they might be available for printing, was immediately irresistible.

Richard Zauft (left) working in the Museum's printing facility with one of his students.

When I first saw the future Museum space in 1997 it was largely empty except for a few type cabinets, type cutting and planning machinery, some palettes of rough-cut wood slabs, and a few boxes of uncut wood pieces. After touring the space with James Donnelly and Jim Van Lanen, I reviewed the proposed floor plan and maquette Donnelly and his design staff had created. The glaring omission that was immediately apparent to me was the absence of a printing facility. After all, I thought, how could a museum devoted to wood type not show how it was used? A printing facility in the Museum seemed essential. Not only would it show Museum visitors how wood type was printed, but it would also provide a means for printers and scholars to access and use the vast collection. And since a central part of the Museum's mission was educational, it seemed a printing facility could also be used for demonstrations, workshops, and a variety of printing projects for the Museum.

Like many letterpress printers, I had acquired a small assortment of wood type over the years but had found it difficult to incorporate the various individual letters and partial fonts into my design work and printing projects in any significant way. The recognition that there were thousands of pieces of wood type, patterns and equipment available to make new wood type, was for me a stunning realization. To avoid the Hamilton Wood Type and Printing Museum becoming a sleepy, dusty collection of wood type that few people would see and even fewer use, it seemed a printing program that brought in letterpress printers, book artists, and graphic designers was necessary.

The Hamilton Museum print shop.

Fortunately it didn't take much persuasion to convince the Historical Society's Board that a printing facility was an important part of telling the Hamilton story. The board already embraced the concept of a 'living museum' dedicated to showing visitors how wood type was made. Printing the wood type was just the next step in this process.

Jim Van Lanen located and acquired several Vandercook proofing presses, and within a few months the space had three presses, over two dozen cabinets filled with assorted wood and metal type, work tables, and drying racks. I donated several dozen cans of ink and Jim got local printers to donate a wide assortment of paper and flat storage cabinets. I overhauled and cleaned the first Vandercook SP-15, now the primary press used at the Museum.

Shortly after my first visit to the future Museum in 1997, I was asked to co-curate an exhibition of Wisconsin book artists with Max Yela, Head of Special Collections at the University of Wisconsin-

ABC Type Specimen Sheet by Richard Zauft.

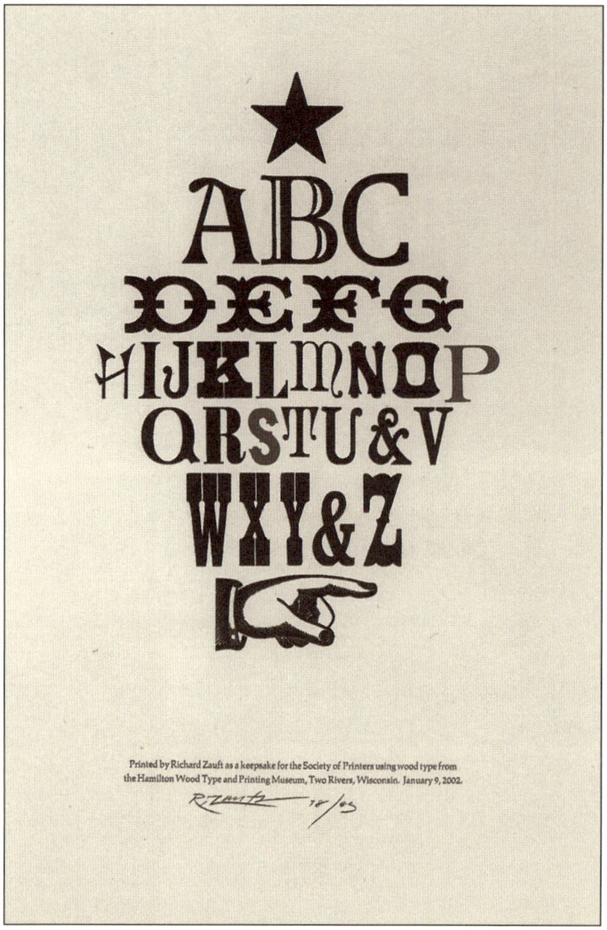

Milwaukee. For that exhibition I designed and printed a commemorative, letterpress printed broadside using wood type from the Hamilton collection. It was the first piece of extensive printing I did using wood type, and the experience was thrilling because the type I used was new. Although cut almost 40 years ago, it had never seen ink. I used a variety of sizes, from 6 line (1 inch) to 6 inch, to print overlapping semi-transparent letterforms as the background to the metal type.
This overprinting yielded a number of secondary and tertiary ink values and unintended abstract shapes to the background design.

This use of wood type to create overlapping color values and abstract shapes inspired Dennis Ichiyama, the first visiting artist at Hamilton, to develop a proposal to conduct a residency at Hamilton during a sabbatical leave from Purdue University.

I first met Dennis Ichiyama at the Mills College Book Arts conference in 1992. We renewed our acquaintance at the Alabama book arts conference in February 1999 when I told Dennis about the Hamilton Museum and the wood type collection. Dennis was immediately interested, having a keen interest in design and typography from his graduate studies at the Yale School of Art and Design and teaching graphic design at Purdue University. For the next few months we exchanged emails about the possibility of him conducting a sabbatical leave at Hamilton in order to print wood type. Dennis attended the Museum opening in May 1999 and shortly afterwards developed his sabbatical leave proposal, which was granted by Purdue, for him to conduct a residency at Hamilton beginning in January 2000. Jim Van Lanen and I convinced the Two Rivers Historical Society Board that a visiting artist program would bring much needed attention and recognition to the Museum, so they agreed to Dennis Ichiyama's proposal to spend several months at the Museum making prints from their collection. Dennis was specifically interested in using wood type to print and overprint using transparent and opaque ink colors to highlight the beauty of the letterforms. Dennis spent the next several weeks familiarizing himself with the collection and developing his ideas. Dennis' first work at Hamilton was to print a series of type specimen sheets of random letterforms, full specimen fonts, and arrangements of ornaments. His subsequent projects included a portfolio of wood type specimens and keepsakes and broadsides for a variety of national and international events.

In the four years since Dennis' sabbatical in 2000, he continues to bring students to Two Rivers to assist him on special projects and give them the opportunity to design and print their own ideas using wood type. Dennis Ichiyama was the first visiting artist at Hamilton, and his sensitivity to the local community, cooperation in working with and utilizing the expertise of the volunteers, and his productivity made his residency a success. Dennis and his students continue to make the trip between Indiana and Two Rivers to print wood type to create new and exciting designs printed in the Museum's printing facility.

After Dennis Ichiyama's residency in early 2000 I began conducting wood type printing workshops, presenting lectures, and organizing exhibitions across the country in order to further introduce audiences to the Museum and its growing collection of wood type, advertising cuts, and printing equipment. Support from the Museum made it possible for me to give lectures and conduct printing workshops at the San Francisco Center for the Book, the Minnesota Center for Book Arts, Indiana University, Wellesley College, the Rochester Institute of Technology, Washington University in St. Louis, and the Society of Printers in Boston. During workshops I bring various pieces of wood type from the Hamilton collection and guide participants to collaboratively design and print broadsides.

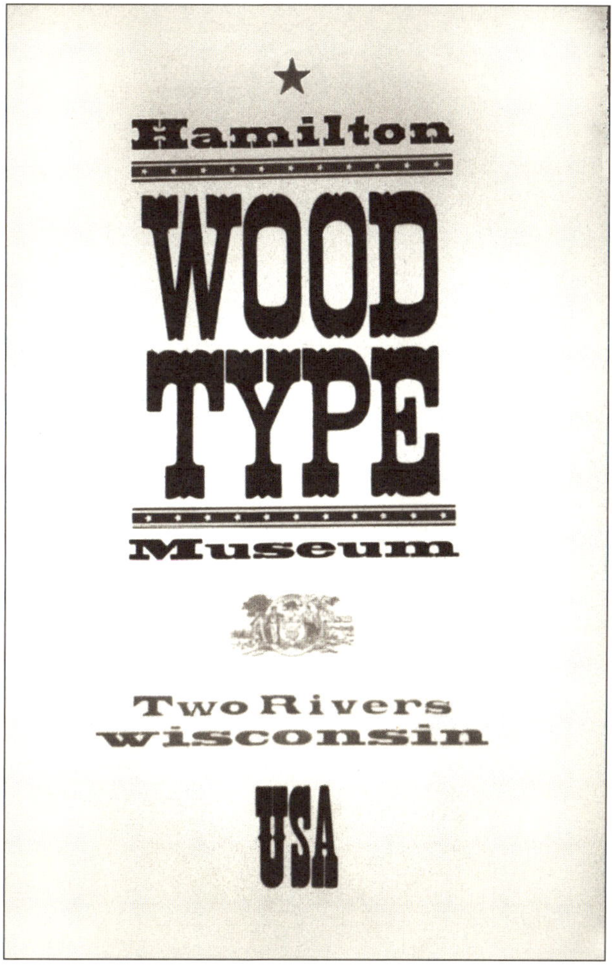

Letterpress broadside by Richard Zauft.

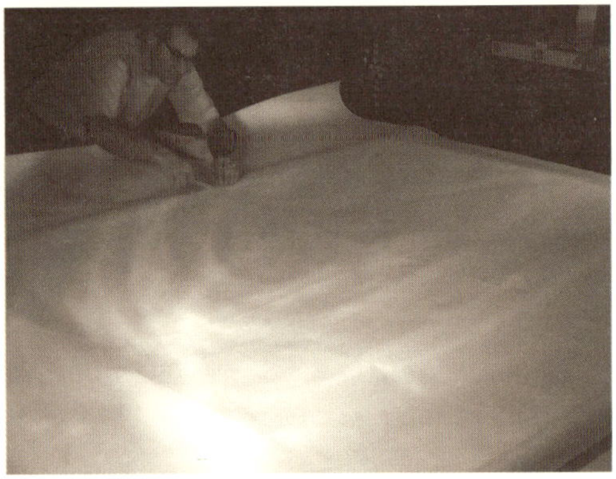

Richard Zauft proofing a large, Tuscan number '2'.

In my lectures I show slides of the museum and its facilities, bring examples of uncut slabs, patterns, and finished letters for the audience to handle and inspect. The biggest "show-stopper" at my lectures is the 81" by 50" number 2 I printed by hand. These lectures and workshops have led to a number of inquiries from collectors, printers, and artists interested in visiting the Museum and using the collection.

In 2002 a number of printers came to Two Rivers to use the collection. Among them was Ruth McGurk of Santa Cruz, California, who visited the museum in the late summer for several intense weeks of printing. Her background as a letterpress book artist gave her an immediate appreciation for wood type. During her four-week residency she combined wood typefaces and advertising cuts to create a series of humorous and satirical broadsides about a range of topics including baseball, rascals and villains, and uppity women. Also visiting that summer were Dan Sjorgen and Jenny Thomas from Denver, Colorado. They focused on the thousands of advertising cuts that the Museum has acquired from local job printers to design and print a number of humorous slogans, broadsides, and postcards. In October the Typocrafters met in Milwaukee for their annual meeting, and spent a day in Two Rivers at the Hamilton Museum printing a collaborative keepsake. Cliff Harvey, from West Virginia, led a weeklong workshop for the Society of Wood Engravers. They created a four-color wood engraving using hard rock Maple wood slabs from the Hamilton collection.

In autumn 2002 Mark Snyder, a professor of art at the Hartford Art School in Hartford, Connecticut, visited to conduct a two-week residency. During his stay he printed over two-dozen broadsides, combing letterforms to create visual puns. Mark is in the process of scanning the letterpress-printed broadsides in order to digitize the letterforms and further manipulate them on the computer before outputting the new designs. Beginning with

Hamilton Offset font by Bill Moran.

Clover Blossom by Ruth McGurk

nineteenth-century wood type letters, Mark is using them to create new letterforms via digital technology in the twenty-first century. This technical adaptation is something James Edward Hamilton could have never imagined.

Susan Merritt, a design professor from San Diego State University, visited the Museum in the summer of 2003. Susan spent a month working with the collection, printing a variety of typographic designs and slogans.

In addition to the small but growing parade of printers and visiting artists, the Museum's collection has also inspired two type design projects. Bill Moran, of Blinc Publishing in St. Paul became aware of the Museum because of his father's printing business and hometown roots in Green Bay, Wisconsin, just a few miles from Two Rivers. As a graphic designer, Bill wanted to design a contemporary typeface to be cut in wood that would appeal to the digital generation of designers. In 2003 he completed the first new typeface designed exclusively for wood in decades. Called "Hamilton Offset," it incorporates features of the offsetting effect that sometimes occurs in letterpress printing. Bill created the patterns utilizing a company that laser cut masonite to create the letterforms for the patterns. These precise patterns were then used to cut the type by hand on the museum's pantograph router. Bill also designed the Museum's web site, which makes Hamilton Offset and three other font faces from the Museum's collection available for purchase in electronic format.

Given to Roam by Mark Snyder.

Carter Latin typeface and the patterns used to route it.

Undoubtedly one of the most significant projects at the Hamilton Wood Type and Printing Museum was the design and creation of a new typeface by Mathew Carter, an internationally recognized type designer. Among the dozens of typefaces he has created or reinterpreted, Carter's best known designs are the typefaces Snell Roundhand (1966), Bell Centennial (1978), Galliard (1978), and Verdana (1996). In early 2002, I approached Mathew Carter about the possibility of designing a typeface in wood for Hamilton, and he responded with guarded enthusiasm. After showing Mathew some wood type from the museum and a few letterpress broadsides I had printed, he confessed he needed to add wood to his repertoire so he could claim that he has designed type in every medium - hot metal, phototypesetting, digital, and wood. In the fall of 2002 Mathew sent me a set of character designs in four word combinations to illustrate his concept for the new face. Matthew Carter's new type design for wood was inspired by the typeface Latin Extended. The wedge-shaped serifs and bold weight is reminiscent of advertising faces of the late nineteenth century. It also embraces a clean, hard-edged boldness that carried a contemporary connotation. Named "Carter Latin," this new face is unique because its outline form allows for both a one-color and a two-color version when interior letterforms are printed inside the "key" outline face.

Jim Van Lanen, Norbert Brylski, and Lloyd Dickensheid at the Hamilton Museum cut the patterns and initial letterforms for Carter Latin in the summer of 2003. Lloyd and Norb are both retired employees of Hamilton who now volunteer at the museum to give tours and demonstrations. Norb was one of the last workers to cut production type for Hamilton in the 1980s before it was discontinued. Norb's continuing primary contribution is cutting typefaces to complete fonts or fill special orders.

In a low-tech procedure consistent with careful handwork at Hamilton, Mathew's black and white letterforms were enlarged, cut out, and glued to thin plywood sheets in order to provide the outlines for hand cutting the patterns on a scroll saw. The interior letterforms were produced the same way. They were slightly enlarged in order to provide the necessary overlap, or "choke" with the outline "key" forms so printing in a second color inside the outline form would have no white hairline borders. The entire pattern making and initial cutting took several attempts to get the registrations properly aligned. The new face was unveiled at the TypeCon 2003 annual conference in Minneapolis in July 2003. Carter Latin is only available through the Hamilton Wood Type and Printing Museum.

The future of this unique collection lies in its ability to be used, studied, and applied to current and future typographic developments. Using the collection to design and print new work, designing new typefaces for production, organizing exhibitions, and investigating the collection for historical and scholarly research assures the museum will be available to future generations of type enthusiasts. The Hamilton Wood Type and Printing Museum invites you to become part of its ongoing contribution to American printing history through your participation.

Education at the Hamilton Museum
by Dennis Ichiyama

Dennis Ichiyama is a professor in the Visual Communications Design Program at Purdue University in West Lafayette, Indiana. As the first Artist-in-Residence at the Hamilton Wood Type and Printing Museum, he has a unique perspective on the educational opportunities it offers. Dennis continues to bring his students to Two Rivers to provide historical context for their studies and to expose them to the creative possibilities offered by wood type and letterpress printing.

Dennis Ichiyama

A museum's role is twofold: preservation and education. For the past four years the Hamilton Wood Type and Printing Museum has preserved a great legacy and history of their city. Education has benefited in many ways, through the numerous visitors ranging from the merely curious and nostalgic, to the scholars, designers and students who have visited and worked with the museum's immense collection of type, decorative borders woodcuts and machinery.

My involvement with the Museum began at the very beginning. On the opening day, May 28, 1999, I was impressed with the collection and excited about the possibility of utilizing the facility for my own research. I began my design residency at the museum in December 1999. I initially introduced the facility to my graduate students in Visual Communications Design with the hope that they would assist me in my research and also turn their attention and studies to the history and development of American wood type. With the completion of several projects – some experimental, some documentary – I returned to the university with the intention of bringing more students to the Museum to experience the process of letterpress printing and wood type.

Over the next four years, many students made the 300-mile road trip from Indiana to Two Rivers. The visits usually lasted one or two weeks and involved one to three students at a time. They were given some material on wood type prior to the visit. A week was enough time to introduce the student to the letterpress method and the collection. We usually worked on several projects. Some were completed during the week, others were finished on later visits. The projects included my research, keepsakes and broadsides produced for a variety of national and international events, and involved completed overprinting.

Letterpress folder and print by Dennis Ichiyama.

Printing History: New Criteria broadside,
University of Reading by Dennis Ichiyama.

Museum identity proposal for the Hamilton Museum by Luiz Navarro de Andrade.

The first student to assist me was Luiz Navarro de Andradea a graduate student from Brazil who was captivated by the charm of Two Rivers and the cadre of knowledgeable volunteers who gave tours and provided personal and historic insights into Hamilton's place in the history of wood type. He designed several pieces to introduce the general public to the Museum and the collection. He also developed a promotional poster and identity proposal for the Museum, as well as a brochure that is currently being used to promote the Museum. His comments generally reflect the feelings most have when they come to the Museum and work with the collection:

Design is almost made instantaneously nowadays. The use of the computer and the hectic marketplace are compressing the process of design to a minimum. When you take time to visit a place like the Hamilton Museum and experience this wonderful process, it helps you to realize what is missing in your own creative process. I left the place with a different idea of typography design and process. I was inspired not only by the museum, but also by the people and their interest in wood type. I have always been fascinated by the process of design not only the final product but how you get there and how ideas are generated. I only spent four days on my first visit but it was obvious to me that it was a fantastic educational opportunity. I experienced the whole process from picking type with my hands, to placing them on the press, mixing the colors, and printing layer by layer. We took time to discuss projects and collaborate on the solutions; it was an experience that changed my career.

In the spring of 2001, I taught a course at Purdue University called "The Contemporary Book" which focused on the production and design of a limited edition book project. The final stage was the production and printing of the limited edition book at the Hamilton Museum. Mariana Wallig Carvalho produced a book entitled *a suadade*, based on the text of a Brazilian folk song.

***a saudade** by Marianna Wallig Carvalho.*

> *My experience at the Hamilton Wood Type and Printing Museum made me realize how important it is to be able to* ***touch*** *type, to see it up close and explore its topography: serifs loops, bowls, tails, and all. At the Hamilton Museum there is so much variety that one could spend the whole time there exploring. But actually setting the type makes you immediately understand the concept.*
> *When I came back from my first visit I was floating on air. I told a friend how my dream job was to get myself a Vandercook press and just print, print, print.*

Another student, Ryan Wilson, describes his experience simply and eloquently:

> *Since many designers have executed most of their work with only a computer, it is hard for students like us to even imagine design of an earlier century. In the future, it will be great to know that at one point in time I worked with wood type, set metal type, and used a printing press. I am very thankful for my experience at Hamilton.*

In summer of 2001, with assistance from R. Jason Ratliff, we produced several limited edition pieces for the American Chemical Society's 225th anniversary, the second volume of wood type specimen sheets from the Museum's collection and a special project based on the myth of Pandora's Box for Chicago's Center for Book and Paper Arts' Epicenter Press housed at Columbia College of Art and Design. The ACS print was based on letters and numbers that were overlaid to create a colorful, celebratory texture of shapes and forms. The specimen sheet project is the second volume undertaken and reproduces the complete alphabet and numerals of some

Pandora's Box *project by Dennis Ichiyama and R. Jason Ratliff.*

Letterpress prints by Dennis Ichiyama.

of the more unique designs in the collection. Pandora's Box was an interpretive design in which the letter forms spell out the title of the publication, but its meaning is concealed by overprinting the letters. Only one of the 150 pieces in the edition spelled out "HOPE" legibly. What is presented here is the final publication version and the Artist Proof print. Ratliff 's comments are particularly poignant and eloquent considering the impact of the computer on his education and training:

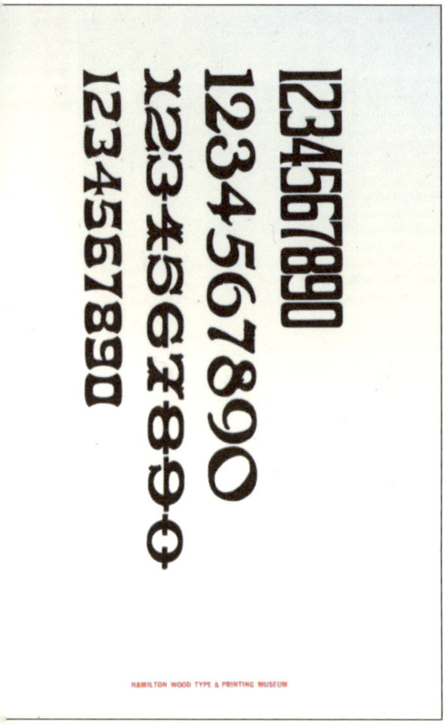

Type specimen sheet from the Hamilton Wood Type Collection.

> *Coming from a generation of mouse-mover and cut-and-paste artists, I left Wisconsin reenergized. I had regained what I thought I lost, inspiration and direction. I feel that there is a certain respect for craft that has been forgotten. The romanticism of how things were once done, touching the actual letter, feeling the paper run through your hands as you feed the press. The feeling you get the moment before you see the printed image, the clean impression, the sharp colors. And no matter how well you plan there is an element of surprise, accidents and mistakes turn into beauty and serendipity. You feel like a kid again, exploring and playing like your first discovery of crayons.*

In the fall of 2002, I was invited to teach a Master Class at the Hartford Art School (HAS) which is part of the University of Hartford (Connecticut). Twelve senior Visual Communication Design (VCD) students were chosen to participate in the course which introduced wood type to their design history and vocabulary. All forms of printing, from letterpress to scanned images and laser printing, were encouraged. The wood type was unique in that it was a collection of large wood type that had broken creating further letterform abstractions. These forms were exciting to behold and a challenge to create with. The previous summer I had done some early experiments was intrigued by the potential of printing the forms using transparent inks and overlapping the imagery. Each student got two wooden forms to print. Over the semester I saw their curiosity turn into excitement which resulted in a remarkable suite of twelve prints in a limited edition.

One of the senior students, Pedja Zkravkovic from Belgrade, Serbia, began a series of studies using transparent inks and colored paper. The result was a marvelous collection of images showing his sensitivity to form, color, and composition.

> *Wood type, fragments of wood type, the weight, texture, cracks, imperfections, the smell of ink, turpentine, alcohol and finger stains, Vandercook press. . . all of these things contribute to the*

> *magical experience of working and creating wood type prints. I remember touching the wooden shape, trying to get a sense of its past, what kind of impressions were left and when. The need to explore the relations between negative and positive space and the effect of color on a design was inspiring.*

VCD HAS 12 was the title chosen for the project which will continue as a personal journey to further explore the potential of color, type, transparency and composition. The time spent at the Hartford Art School was an opportunity to share in the individual visions and exploration of twelve talented students.

The youngest visitor to the Museum was Daniel Heiss, a sophomore in the Division of Art and Design who accompanied me to the museum to continue several projects and begin a few new ones. His reactions to the town, the Museum, the volunteers, and the experience were similar to previous students. However, each one brings their own ideas and visions which have given them an opportunity to learn a new printing process, work with a type fashioned out of beautiful wood, learn about the interaction between color and paper, and participate in a history and tradition that will not only expand but extend their knowledge of typography, design, and communication.

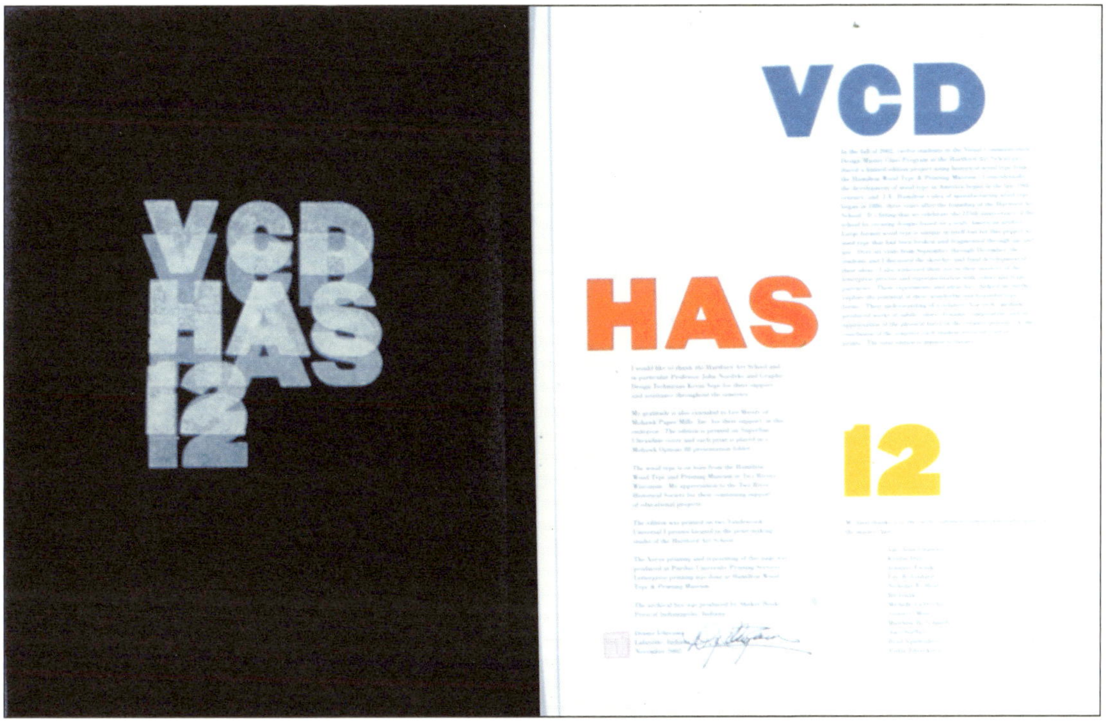

Hartford Art School portfolio project.

The Hamilton Wood Type Museum is a unique place of history preserved for future generations. The building is a safe haven for all wood type, metal type, and related machinery. Each individual piece of wood type is just waiting to be looked at, touched, and aching for ink. The many different fonts housed in the museum are not just wood type; each piece is an opportunity for a creative mind to be inspired.

On the fourth anniversary (May 2003) of the Museum our visit was particularly eventful when Ryan Wade, who had recently graduated from Purdue University, had the unique opportunity to print a keepsake using type created by Darius Wells.

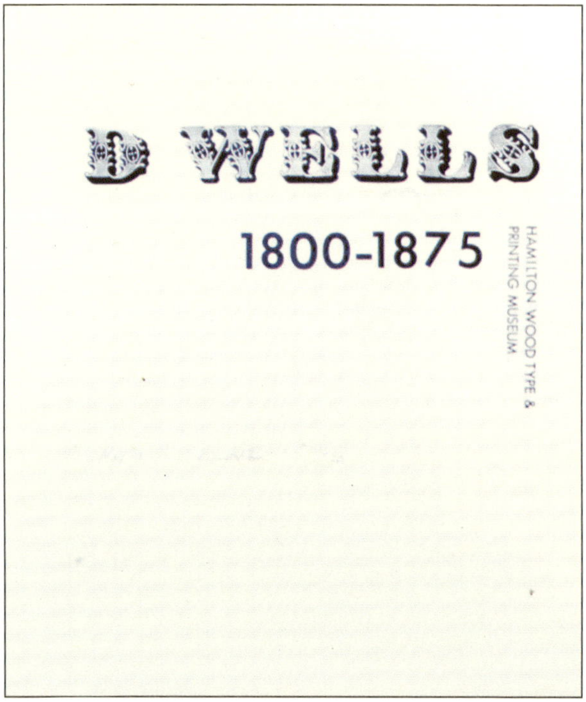

D. Wells* broadside *by Ryan Wade.

The opportunity to bring back life to over a century-old Darius Wells wood type provides a sense of community to a great pioneer of graphic design and typography. Having done research on the history of type, I never thought I would be working with Wells' wood type on a Vandercook press at the former Hamilton Manufacturing Company. This priceless experience was not only extremely educational, but provided lasting memories which I will be able to carry with me the rest of my life.

Since 1999, over a dozen students from Purdue have visited the Museum. The students came from Brazil, China, Germany and the United States and their visits range from four days to two weeks. No matter the length of stay, their excitement learning a "new process" and working with historic materials fill the hours and make for a most rewarding experience.

Throughout my affiliation with the Hamilton we've had the generous and warm hospitality of the volunteers who make the experience real and alive. Two members deserve special recognition for their interest in the students and their projects. Dan Petrashak had been part of the Museum since my arrival in 1999. His recent passing has created a irreplaceable void. Norbert Brylski is another volunteer who operates the museum's pantograph and demonstrates the craft of cutting and finishing each wood type. Every student has had the thrill of cutting a letter which has become a cherished souvenir of their visit. These volunteers keep the art, craft and history of wood type alive and relevant.

Our work at Hamilton could not continue without the generous support of the paper industry. Over the past four years Mohawk Paper Mills, Inc. has been a supporter of the Museum and its educational mission. Its commitment has enabled me to continue to provide a unique experience for our students and continue our research.

The mission of the Hamilton Museum has been fulfilled in the examples and experiences of these students. As a teacher and designer I applaud and give thanks to James Van Lanen Sr. for his vision and to the Two Rivers Historical Society board for their continual support of the educational mission of the Hamilton Museum. The Museum will continue to play a major role in education and will ensure that future generations can experience the sheer joy of handling a unique piece of American printing history.

Acknowledgements

The authors and Hamilton Wood Type and Printing Museum wish to thank the following individuals and institutions for their cooperation and generosity.

Janet Abrams
Terry Bosch
J. Michael Brown
Mary Helen Kelly
Bernie and Pat Moran
James P. Moran
Mary Jo Pauly
Dave DeRosier
WI Nationwide
Dennis Ichiyama
Richard Zauft
Jeff Waldvogel
Mohawk Paper Company
Fisher Hamilton
Mardel Doubek
Jeff Cisler
Dan Petrashek
Wilfred Brocher
Norbert Brylski
Lloyd Dickenshied
Mabel Stechmesser
Darlene Guehlstorf
Gene Petri
Francis LeClair
Herb Bunke
Pat Christel
Joe Jebavy
Dan Solo
Catherine and James Donnelly
James Van Lanen Sr.
George Cope
Ken Bastar
Orrin Rohde
Bob Mueller
Clem Leonhard
Joe Kupsch
Russell Goedjen
Dean Konop
Don Konop
Gerald Schultz
Glenn Leege
Francis Lyons
Craig Powell
Gary Bernstein/Neenah Paper

HAMILTON
WOOD TYPE